Who Needs To Sleep Anyway?

How Animals Sleep

Who Needs To Sleep Anyway?

How Animals Sleep

by Dr. Colin Shapiro

illustrated by Sari O'Sullivan

Published in March, 1996 by Black Moss Press,
2450 Byng Road,
Windsor, Ontario, Canada,
N8W 3E8

with the assistance of the Canada Council
and the Ontario Arts Council.

Black Moss books are distributed in Canada by:
Firefly Books,
Firefly Books
3680 Victoria Park Avenue
Willowdale, Ontario
M2H 3K1

and in the United States by:
Firefly Books (U.S.) Inc.
P.O. Box 1338
Ellicott Station
Buffalo, New York
14205

Cataloguing in Publication data

Who needs to sleep anyway? how animals sleep
ISBN 0-88753-288-8 (bound)
ISBN 0-88753-281-0 (pbk)

1. Sleep behavior in animals--Juvenile literature.
I. O'Sullivan, Sari   II. Title.
Ql755.3.S43 1996   j591.5   C96-900263-7

For all those children who want to learn more about sleep

This book is a true story about who sleeps. It started when my daughter Mahla asked me: "Why do I need to go to sleep? Who needs to sleep anyway?"

Well, of course, that is really two questions rolled into one and I decided that the easier question to answer was "who needs to sleep anyway?" And this is what we talked about.

“Why do I need to go to sleep? Who needs to sleep anyway?” Mahla asked me one night after she put on her pajamas and didn’t want to go to bed.

“Lie down in your bed, Mahla,” I said, “and I will tell the story of who sleeps and how. But first, you must promise me that you will close your eyes when I finish and go to sleep yourself.”

Mahla lay her head on her pillow, her teddy clutched in her arms. “Okay. It’s a deal,” she said.

I lay down beside her. “You’ve seen Pandy and Themba sleep during the day,” I began.

“But they’re dogs. And they spend all night guarding the house in case robbers come.”

“Right,” I said. “Dogs take little naps all day long and don’t sleep through the night. That’s their pattern and that’s their job.”

“Maybe they only have little naps because they’re inside the house,” Mahla said.

“But you’ve seen them snoozing in the sun or in the back seat of the car—anywhere they feel like it and they feel comfortable. Some animals sleep only once a day and have one long snooze, but others sleep only a little, then wake up for a while before going back to sleep. Newborn human babies do that, and sometimes very old people do it too. The medical name for this is *polyphastic sleeping*, but we call it *cat-napping* or *dog-napping* because that’s what cats and dogs do.”

“Sometimes Pandy sleeps on her back, just like we do. And Themba too. With all their legs in the air.” Mahla wriggled out of the covers to demonstrate.

“But they don’t snore,” I was quick to point out, easing her back down on her pillow.

"Yes, but sometimes they whisper and sort of bark in their sleep. And their paws move too.

"That's because they're dreaming."

"Like I do? Every time they lie down? What do they dream about? Do all animals dream? And who sleeps the most? And does it matter if they sleep inside or outside?" Mahla sat up in bed.

"That's a lot of questions!" I said. "Lie down. I can't tell you the story if you're not lying down."

Once she had settled down again, I said, "Well, where do we start? Animals, just like children and mothers and fathers, need sleep. And they dream just like you, although I don't think they dream about the same things. Now let's see, what other animals do we know?"

"Suzette's cat, Jamima! She's always sleeping. Indoors and out."

"That's right," I said. "Jamima often sleeps outside when she's not sleeping in the house. Cats nearly hold the record for sleeping."

"Jamima holds the world record for sleeping? I can hardly wait to tell Suzette," Mahla said.

"But not right now," I said.

"Why does she wiggle her whiskers when she sleeps? Is it the same as Pandy and Themba moving their paws?" Mahla asked.

"Now, that's a good question, and you're right," I said.

"When you dream, your eyes move around a lot under their lids. That's when you "see" the pictures of the dream, like a movie inside your head. Animals' eyes move in the same way. That's how we can tell they're dreaming. Our dogs are probably dreaming about playing with you or chasing a cat. When cats dream, their whiskers twitch as if they're sensing something.

Sari

"Sometimes, a cat sits like a sphinx, bolt upright with its eyes closed. When it starts to dream, the muscles in its neck relax and its head sags down. If the whiskers are trembling and the head is down, the cat is dreaming. And if it's a really exciting dream, maybe about catching a mouse, its tail and paws will start to twitch too," I said.

"And what happens when I wake her up when she's dreaming?"

"Why would you want to do that?" I asked.

"If it's a bad dream..."

"She'd wake herself up, just like you do. Now, that's enough for tonight. It's time for you to go to sleep," I said.

"But you haven't finished." Mahla grabbed my sleeve before I could get up. "You promised you'd tell me how *all* animals sleep, and you've only talked about dogs and cats. What about wild animals?"

"Okay. I'll tell you a few more stories. But then, you must promise to go to sleep," I said.

"I will."

"Do you remember when we went to Hawaii?" I continued.

"We saw whales!" Mahla bounced up and down. "They came right up to the side of the boat. And they blew water through holes in their heads."

"That's right. That's how they breathe."

Mahla laughed. "I can only use my nose."

I tweaked it. "It's a pretty good nose for a human child. Now, where do you think whales sleep?"

"Well, if they sleep", said Mahla, "I suppose they sleep on the bottom of the sea."

"No, silly," said I. "They'd drown if they did that. A whale is a mammal, just like you and all mammals have to breathe air whether they're asleep or awake."

"Some kids can't." Mahla pointed out. "Sometimes my sister gets asthma in the middle of the night and she can't breathe. She wakes everybody up."

"That's another story for another time," I said. "Don't try to distract me. When whales sleep, they lie just below the surface of the water with their heads sticking out just far enough so they can breathe through their blow holes."

"That would look pretty funny—an ocean full of sleeping whales like water lilies on a pond. I remember dolphins too. Do they sleep like that?" Mahla asked.

"They're almost as clever as you are," I said, tickling her only a little. "They have a really tricky trick. They sleep with one half of their brain part of the time and then, they switch over and sleep with the other half, so when they wake up, they feel grand. When they're sleeping like that, they're always moving around in the water, not just floating like whales. Otherwise, they would drown."

"How did you figure that out?"

"I didn't. Some Russian scientists recorded the sleep of dolphins with machines and electrodes. That's one of the ways we can tell if animals are sleeping. We can use these to measure how active your brain is and so we can tell if animal is really sleeping," I said.

"Did you ever experiment on dolphins, Daddy?"

"Not on dolphins, but I did do some experiments on fish."

"Fish don't sleep," Mahla insisted. "They swim all the time."

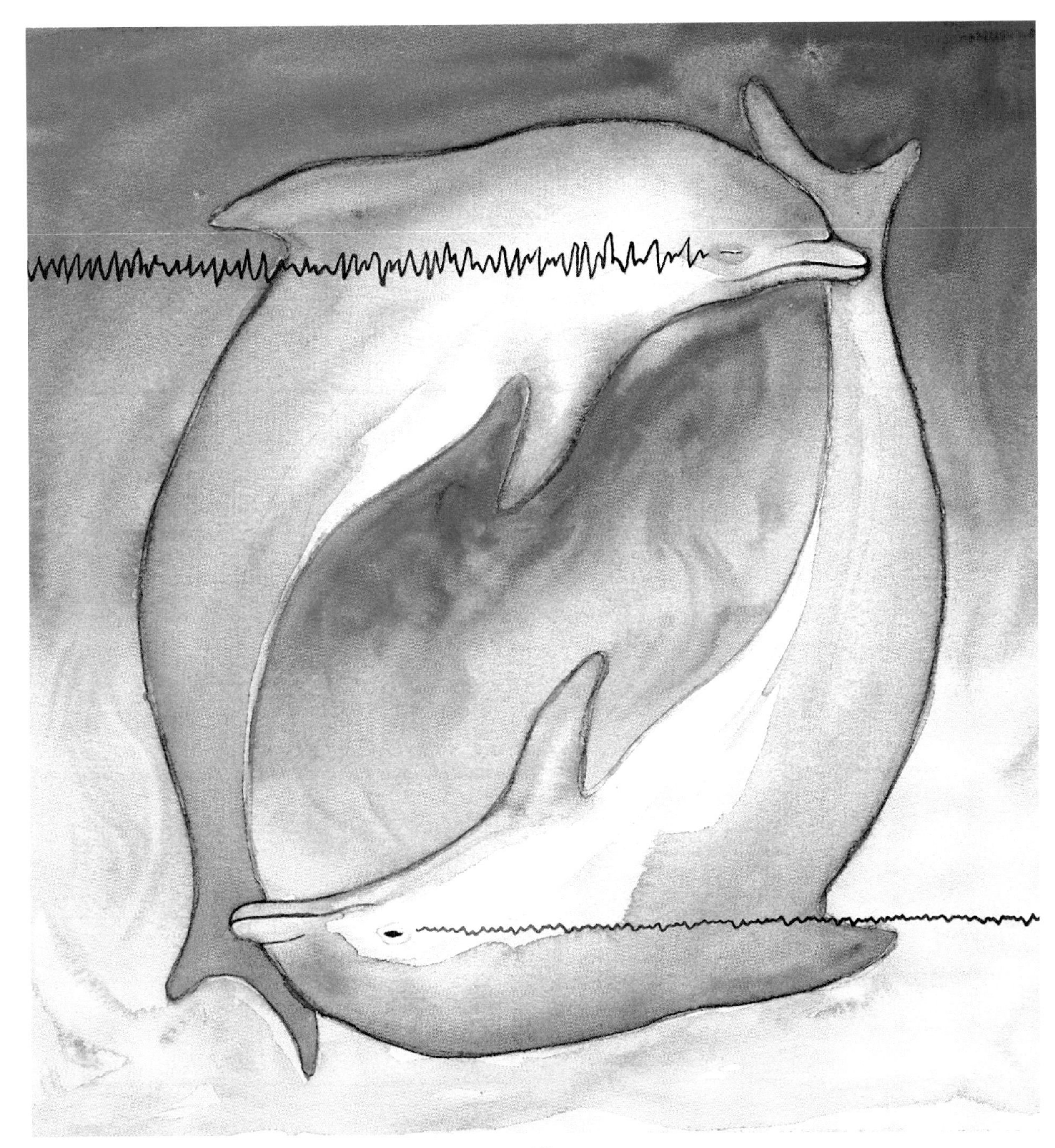

"All living things sleep," I answered. "Even fish. But because they're not mammals like dolphins and whales, they don't breathe air. So they can sleep under water, even down at the bottom of the ocean.

"When I was in Africa, I had a chance to study whether fish slept. Some friends of mine were trying to figure out if fish had memories. They put floats on a bunch of *Tilapia mossambica*."

"What's that?" Mahla giggled. "Tilla mossa what?"

"We call them guppies for short," I said.

"Like Erin has in her fish tank?"

"Exactly the same. Anyway, when my friends put the floats on, at first the guppies would swim upside down, but after a while, they learned how to swim with them. The experiment measured how long it took the fish to learn to swim the right way up. One night, they came to the lab at three in the morning. Everything was dark and still. They turned on the lights. In every row of tanks, all the fish were lying at the bottom. My friends wondered if they were sleeping.

"They asked me to find out so I put some of the fish in a room without windows. I left the lights switched on all night and turned them all off, except for one little light, during the day so I could see what the fish did in the dark. That way, I wouldn't have to come to the lab in the middle of the night. I like to sleep too," I said.

Mahla didn't get the hint. "What happened to the fish?"

"Well, they all slept on the bottom of the tank. And when they slept, they slept so deeply they didn't know what was happening around them."

"How did you know that?" Mahla asked.

"When fish are swimming around and you throw bread into the tanks,

Sari/94

they gobble it up very quickly. If they're just resting on the bottom of the tank, they'll come to the top to get the bread. But if they're sleeping, they'll ignore the food and it sinks down to the bottom," I told Mahla.

"If someone comes into your room when you're sleeping and whispers your name, you don't hear them. But if you're awake, you can hear even the smallest whisper because you're alert then to everything going on around you. It's the same as the fish and other animals," I said.

"And there's another funny thing. Guppies have stripes along their bodies. When they wake up, they swim backwards for a few minutes and when they do, the stripes change from going along their bodies to going up and down. It's like the fish are changing from pajamas into clothes."

"Why?" Mahla asked.

"The light changes the pigment in their skin and it makes the stripes change direction. And once the stripes change, even though the backward swimming doesn't last very long, they stay changed all day."

Mahla held her hand up to the light. "My skin doesn't change colour."

"You're not a guppy," I said. "Now, if you don't have any more questions..."

"But I do." Mahla sat up again. "We've hardly started. What about crocodiles?" She began to sing:

Boo hoo hoo, I'm a lonely crock
Lying all day on a sunny rock
I want a friend, oh! with all my might
But nobody likes my appetite.

I laughed, but then spoke in a pretend stern voice. "Do you want to sing songs or hear stories? Because it's not the time for singing now."

Sari

"Stories," Mahla said. "Do crocodiles sleep?"

"They do. But that was very hard to prove. Usually when people are sleeping, they have a special way of lying down. For instance, you cuddle into a ball. And you don't know what's going on around you, just like the fish that didn't respond to food. When you're sleeping, you don't notice if someone touches you, unless they touch you quite hard. But if you are awake, you'll notice it pretty quickly," I said.

"And, if you're sleeping, you have to be able to wake up. If you can't wake up, you'd be unconscious and it would just look like you were sleeping. We look for the same signs in animals to tell whether they're asleep.

"Well, there was a man named Mr. Flanigan who was very interested in crocodiles. He studied crocodiles in his laboratory to see if they sleep. He was partly convinced that they never slept because they never seemed to relax or do any of these things which tell if an animal or person is sleeping.

"You see, crocodiles always look as if they're sleeping, for one thing. But they also always seem to be alert. And they don't curl up in a ball like some animals do or show any other sleep patterns we're used to seeing. Then, one Sunday, he came into his lab and found all the crocodiles fast asleep. He was very puzzled, because the crocodiles couldn't know it was Sunday, though people sometimes nap on Sunday afternoons," I said.

"You do," Mahla said.

I kept right on with the story. "He figured out that nobody used the elevator in the building where the laboratory was all weekend. Crocodiles are very sensitive to movement and when the elevator was being used, the whole building would rock a little bit. So, all week long, they stayed awake. But once everyone left and it was quiet and still, they all fell asleep."

Mahla's eyes were closed. I started to ease myself off the bed. Then she

asked, "Do birds sleep?"

"I know one who doesn't," I said.

"Really, Daddy, do birds sleep?"

"Not as much as people or cats or dogs, but more than some other animals. Eagles and vultures sleep a lot and they dream, too," I said.

"Roosters wake everybody up. Cock-a-doodle-do."

"That's right. Chickens sleep, sometimes with their heads under their wings. But the rooster wakes them up, too."

"Why does he do it? Is he mean?" Mahla asked.

"No. It's just his nature. Animals all have different amounts of sleep that's right for them and they all have times that are right for them to sleep.

"Some birds, like owls, and some animals like to sleep during the day so they can hunt all night. Others go to sleep when the sun goes down and wake up when it rises. Even people sleep at different times and different amounts of time. Children, for instance, should go to bed early to get their best sleep," I said.

"Does that mean older kids should go to bed late?"

"Not necessarily" I said. That's

too smart a question and it's a story for your older sister Zoe when she becomes a teenager and anyway I have some other interesting things to tell you before we wrap it up and wrap you up.

"Is that why you stay up so late, because you're old?"

"Let me tell you about some other creatures," I said quickly. "Bats sleep upside down, hanging from a tree branch or from the tops of caves."

"What happens when they dream? Do they let go?" Mahla asked.

"No silly," I said. "They'd fall and bump their heads. Each animal seems to know how to sleep in the right place for it. A leopard can sleep on a tree limb without falling but you couldn't do that and whales can sleep floating in the water. We think all animals have some inner sense of balance that allows them to sleep in the place they like best."

"Do penguins dream?" Mahla asked.

"Yes, but I don't know what about."

"Hot soup!" Mahla suggested.

"Could be. Anyway, they sleep in a gang. They get into a big circle when the wind is blowing and it's icy cold. Those with their backs to the wind don't get much sleep but those protected by the others are warmer and can sleep standing up. Then the whole flock turns, so each bird gets a chance to sleep."

"I can sleep standing up," Mahla started to get to her feet on top of the bed.

"No you can't!" We started to wrestle and giggle.

"What's all that racket?" Mahla's mom called from the hall. "I thought you were telling Mahla a bed-time story."

"I am," I answered. I tucked Mahla back in. " And I'm done. It's time for you to keep your part of the bargain and go to sleep."

"But who sleeps the most?"

"I'll go right down to my desk and make you a list. And when you wake up in the morning, it will be right here on your dresser," I said.

"All right." Mahla curled into a ball, her fist against her cheek. I turned off the light. "Good night, sleep tight," I whispered. "Don't let the bed-bugs bite."

"Bugs, what about bugs? Do they sleep? Do spiders and ants and caterpillars and bees...and why do they sleep?"

“Enough, enough!” I held my hands to my ears. “I know that you like bug stories and I will tell you one more story about them, but then it really is time to go to sleep.

“Just like you, even bugs and spiders and scorpions seem to sleep some of the time. Bees rest quite a bit because when it is daytime, they are going to be very busy. There’s a famous description of how bees and animals and even people act during the day called the *rest-activity cycle*. It means that all through the day, people rest for a while and then are very active for a while. About every 90 minutes, grownups feel like taking a bit of a rest. It's shorter in kids and it's even shorter in bees and bugs and insects," I said.

“But every living thing needs rest and every living thing needs sleep. I’ll tell you why tomorrow night. But now, if you want that list before I go to bed, it's time for you...”

“I know,” she said, “to go to sleep. And I know that I’ll dream the way that penguins do.” And with that, she turned her back and closed her eyes and went to sleep.

I went to my study and made a list of how long some animals sleep every day and wrote down the time people sleep on average next to it. If you want to see it, just turn the page.

SLEEP DURATION FOR ANIMALS

Animals	Total sleep time every 24 hours	Humans at different ages
Little brown bat	20 hours	
North American opossum	19 hours	
Giant armadillo	18 hours	Before birth
Nine-banded armadillo	17 hours	
Squirrel	16 hours	Newborn
African python	15 hours	
Mongolian gerbil	15 hours	
Cockroach	14 hours	3 - 5 months
Domestic cat	13 hours	6 - 23 months
Domestic dog	12 hours	2 - 3 years old
White leghorn chicken	11 hours	3 - 5 years old
Emperor penguin	10 hours	
Jaguar	10 hours	5 - 9 years old
Parakeet	9 hours	10 - 13 years old
Great tit (a bird)	8 hours	14 - 18 years old
Indus dolphin	7 hours	33 - 45 years old

SLEEP DURATION FOR ANIMALS

Animals	Total sleep time every 24 hours	Humans at different ages
Guppy	6 hours	50 years old
Goose	6 hours	
Falcon	5 hours	90 years old
Wild turkey	4 hours	
Crocodile	3 hours	
Horse	2 hours	
Red kangaroo	1 hour	
Tortoise	less than 1 hour